C000145517

20 FUN FACTS ABOUT
FAMOUS CANALS AND SEAWAYS

BY CAITIE McANENEY

Gareth Stevens
PUBLISHING

Please visit our website, www.garethstevens.com. For a free color catalog of all our high-quality books, call toll free 1-800-542-2595 or fax 1-877-542-2596.

Library of Congress Cataloging-in-Publication Data

Names: McAneney, Caitie, author.
Title: 20 fun facts about famous canals and seaways / Caitie McAneney.
Other titles: Twenty fun facts about famous canals and seaways
Description: New York : Gareth Stevens Publishing, 2020. | Series: Fun fact
  file : engineering marvels | Includes index.
Identifiers: LCCN 2019016320| ISBN 9781538246542 (pbk.) | ISBN 9781538246566
  (library bound) | ISBN 9781538246559 (6 pack)
Subjects: LCSH: Waterways–Miscellanea–Juvenile literature. |
  Canals–Miscellanea–Juvenile literature.
Classification: LCC TC745 .M39 2020 | DDC 627/.13–dc23
LC record available at https://lccn.loc.gov/2019016320

First Edition

Published in 2020 by
Gareth Stevens Publishing
111 East 14th Street, Suite 349
New York, NY 10003

Copyright © 2020 Gareth Stevens Publishing

Designer: Sarah Liddell
Editor: Therese Shea

Photo credits: Cover, p. 1 (main) Leonard Zhukovsky/Shutterstock.com; file folder used throughout David Smart/Shutterstock.com; binder clip used throughout luckyraccoon/ Shutterstock.com; wood grain background used throughout ARENA Creative/ Shutterstock.com; p. 5 S-F/Shutterstock.com; p. 6 zhao jiankang/Shutterstock.com; p. 7 Bloomberg/Contributor/Bloomberg/Getty Images; p. 8 HansFree/Shutterstock.com; p. 9 TheZAStudio/Shutterstock.com; p. 10 Stock Montage/Contributor/Archive Photos/ Getty Images; p. 11 WitR/Shutterstock.com; p. 12 Marilyn Angel Wynn/Nativestock/ Getty Images Plus/Getty Images; p. 13 N McIntosh/Shutterstock.com; p. 14 Jaro68/ Shutterstock.com; pp. 15, 18 Andrei Minsk/Shutterstock.com; pp. 16, 17 Everett Historical/ Shutterstock.com; p. 19 DEA/BIBLIOTECA AMBROSIANA/Contributor/De Agostini/ Getty Images; pp. 20, 22 Russ Henil/Shutterstock.com; p. 21 RLS Photo/Shutterstock.com; p. 23 Harvey Meston/Staff/Archive Photos/Getty Images; p. 24 Florida Stock/ Shutterstock.com; p. 25 Lil' Digital Shop/Shutterstock.com; p. 26 VEK Australia/ Shutterstock.com; p. 27 zstock/Shutterstock.com; p. 29 Juriah Mosin/Shutterstock.com.

Printed in the United States of America

Some of the images in this book illustrate individuals who are models. The depictions do not imply actual situations or events.

CPSIA compliance information: Batch #CW20GS: For further information contact Gareth Stevens, New York, New York at 1-800-542-2595.

CONTENTS

Words in the glossary appear in **bold** type the first time they are used in the text.

WATERWAYS AROUND THE WORLD

For thousands of years, people have used water as a source of **transportation**. Waterways can help people as well as goods get from place to place. But what happens when there's no waterway to get where you want to go? You have to build one!

Man-made waterways have changed the course of **civilizations** and the way our world looks. Digging some of these channels took many years and thousands of workers. However, these marvels, or wonders, were worth the work!

In some places, people still use canals for everyday travel, such as in the Netherlands. A canal is a man-made waterway for travel or for moving water from place to place.

5

THE LONGEST — AND OLDEST!

THE GRAND CANAL IS THE LONGEST MAN-MADE WATERWAY.

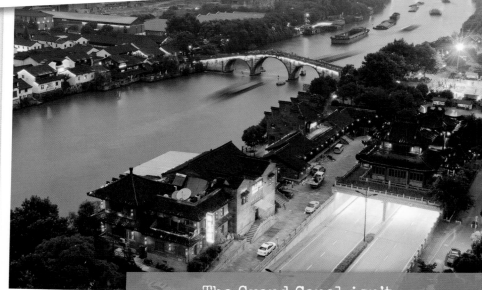

The Grand Canal in China connects the cities of Beijing and Hangzhou and links the Yellow and Yangtze Rivers. It's about 1,100 miles (1,770 km) long—or about the distance from New York City to Kansas City, Missouri!

The Grand Canal isn't entirely a canal. It's not all man-made. Rivers make up some of its pathways.

The Grand Canal has helped carry goods through China for thousands of years!

THE ANCIENT CHINESE STARTED THE GRAND CANAL ABOUT 2,500 YEARS AGO!

The oldest part of the Grand Canal was likely first built around the 400s BC and was rebuilt in AD 607. It's the oldest canal in use. Different parts were added to it over the years.

7

FUN FACT: 3

THE EARLIEST KNOWN CANAL SYSTEM WAS BUILT MORE THAN 4,000 YEARS AGO—BY THE FIRST COMPLEX CIVILIZATION.

The ancient Sumerians were the oldest advanced civilization. They settled in Mesopotamia, today's Iraq, where they built canals to bring water from the Tigris and Euphrates Rivers to their farms.

Ancient Sumerians built great cities and created the first system of writing. This picture shows Sumerian cuneiform, which was their kind of writing.

Without **technology**, some ancient ruins, such as this temple of the Maya people, might be lost forever.

A NEW MAPPING METHOD FOUND AN ANCIENT MAYA CITY IN GUATEMALA — INCLUDING CANALS.

Technology has made it easier to see into the jungle of Central America. Scientists found the remains of ancient Maya farm fields, **pyramids**, and **irrigation** canals, which may date back 1,200 years!

EGYPTIAN CANALS

MOST CANALS HAVE LOCKS—BUT NOT THE SUEZ CANAL.

Locks raise or lower ships when water levels change, so ships don't get stuck. The Suez Canal connects the Mediterranean Sea and the Gulf of Suez, which are about the same water level. Locks aren't needed.

This picture shows workers dredging, or digging out, the Suez Canal. Boats could travel between the Mediterranean Sea and the Red Sea after its completion.

10

In ancient times, pharaohs, the rulers of Egypt, ordered many great **engineering** projects, including huge pyramids and long canals.

AN EARLY MAN-MADE WATERWAY NEAR THE SUEZ CANAL — THE CANAL OF THE PHARAOHS — MAY DATE BACK TO AROUND 1850 BC!

The Suez Canal was finished in 1869. However, people had been trying to connect waterways in that area for centuries. Ancient Egyptians built the Canal of the Pharaohs to link the Nile River to the Red Sea.

11

ANCIENT AMERICAN CANALS

FUN FACT: 7

CANALS FROM AROUND AD 250 WERE FOUND IN FLORIDA.

The Ortona people dug canals from Lake Okeechobee to the Gulf of Mexico using shells and wood! Today, these canals look like dents in the ground. They were once about 20 feet (6 m) wide and 4 feet (1.2 m) deep!

The Ortona traveled in dugout canoes that looked somewhat like this.

There are still reminders of the Hohokam people in central and southern Arizona like this petroglyph, or rock carving. They lived in the area from around AD 200 to 1400.

ANCIENT CANALS IN ARIZONA WATERED FARMS IN A VERY DRY PLACE.

The Hohokam people of today's Arizona created a complex irrigation system. Their canals were about 12 feet (3.7 m) deep and more than 150 miles (241 km) long. They carried water from the Salt River inland to farms.

A CITY OF CANALS

VENICE, ITALY, IS MADE UP OF 118 ISLANDS—CONNECTED BY MORE THAN 200 CANALS!

Founded on islands in the early 600s, people built canals to create Venice. Today, canals are still used for transportation. People travel around in different kinds of boats, such as gondolas.

This is the Grand Canal in Venice, one of the most famous canals in the world. Visitors sometimes travel in gondolas, shown here.

VISIT A CANAL CITY!

AMSTERDAM, NETHERLANDS

STOCKHOLM, SWEDEN

ST. PETERSBURG, RUSSIA

BRUGES, BELGIUM

VENICE, ITALY

NEWPORT BEACH, CALIFORNIA, USA

YANGZHOU, CHINA

ALAPPUZHA, INDIA

TIGRE, ARGENTINA

BANGKOK, THAILAND

A HISTORIC CANAL

THE ERIE CANAL HELPED THE NORTH WIN THE AMERICAN CIVIL WAR.

The Erie Canal stretches across New York State. It made the **economy** of the Northeast boom and linked New York to the Midwest. When the Civil War began in 1861, the Midwest supported the North partly because of this connection.

The Erie Canal allowed boats full of grain and other goods to travel between the Great Lakes and the Atlantic Ocean.

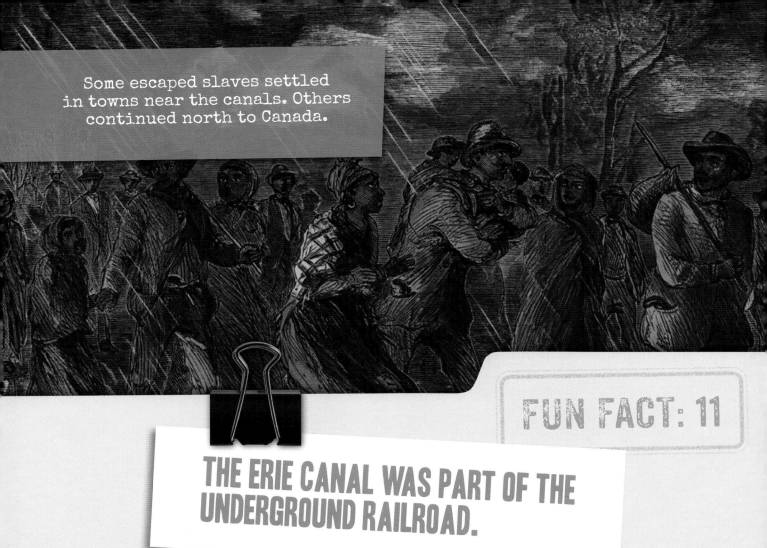

Some escaped slaves settled in towns near the canals. Others continued north to Canada.

THE ERIE CANAL WAS PART OF THE UNDERGROUND RAILROAD.

Before the Civil War, escaped slaves from the South traveled to the North on the Underground Railroad, a system of secret paths, safe houses, and helpful guides. The canals of New York, including the Erie Canal, were routes, or paths, for slaves.

FUN FACT: 12

SHIPS TRAVELING BETWEEN THE EASTERN AND WESTERN UNITED STATES HAD TO GO AROUND SOUTH AMERICA—UNTIL THE PANAMA CANAL.

Opened in 1914, the Panama Canal connected the Atlantic and Pacific Oceans. It shortened the journey for ships traveling between the eastern and western US coasts by about 9,200 miles (14,806 km)!

NORTH AMERICA

UNITED STATES

ATLANTIC OCEAN

PACIFIC OCEAN

PANAMA CANAL

SOUTH AMERICA

This map shows the location of the Panama Canal.

The United States oversaw the construction of the Panama Canal, which Panama now controls.

BETWEEN 1904 AND 1913, MORE THAN 56,000 PEOPLE WORKED ON THE PANAMA CANAL!

Sadly, more than 5,000 workers died during construction of the Panama Canal. Many were killed by yellow fever and malaria—illnesses spread by bugs called mosquitoes.

A GREAT SEAWAY

FUN FACT: 14

IT WOULD TAKE MORE THAN 24 DAYS TO WALK THE WHOLE GREAT LAKES–ST. LAWRENCE SEAWAY!

The Great Lakes–St. Lawrence Seaway connects the Great Lakes to the Atlantic Ocean. The St. Lawrence Seaway part is a river with two locks in the United States and 13 locks in Canada.

The Great Lakes–St. Lawrence Seaway is more than 2,340 miles (3,766 km) long!

The St. Lawrence Seaway portion, or part, stretches from Montreal, Canada, to Lake Erie.

FUN FACT: 15

MORE THAN 200 MILLION TONS (181 MILLION MT) OF GOODS ARE SHIPPED EACH YEAR ON THE GREAT LAKES-ST. LAWRENCE SEAWAY!

Americans and Canadians ship grains, coal, iron, salt, and nearly every other kind of good on the Great Lakes–St. Lawrence Seaway. The weight of all these is more than the weight of 570 Empire State Buildings!

This is a lock on the St. Lawrence Seaway in Ontario, Canada. It's just one of the locks that lifts ships as they travel up the Niagara Escarpment.

FUN FACT: 16

EIGHT LOCKS ON THE ST. LAWRENCE SEAWAY LIFT SHIPS 326 FEET (99 M) OVER THE NIAGARA ESCARPMENT!

An escarpment is a long slope or cliff separating two flat areas of land. The Niagara Escarpment is famous for Niagara Falls. The St. Lawrence Seaway would be impossible for boat travel without its locks.

A SHIP AS LONG AS TWO FOOTBALL FIELDS CAN FIT IN A LOCK ON THE GREAT LAKES-ST. LAWRENCE SEAWAY.

The Great Lakes–St. Lawrence Seaway is important to the economies of both Canada and the United States.

Each lock on the seaway is 766 feet (234 m) long, 80 feet (24 m) wide, and about 30 feet (9 m) deep. It takes a lot of water to fill a space this big—about 24 million gallons (91 million l)!

23

INTRACOASTAL WATERWAY

THE INTRACOASTAL WATERWAY IS THE LENGTH OF ABOUT 114 MARATHON RUNS!

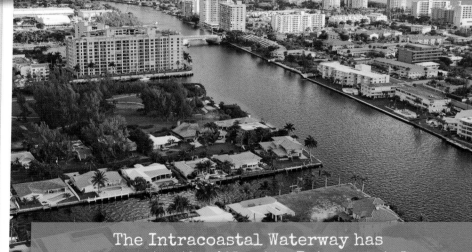

The Intracoastal Waterway shipping route stretches about 3,000 miles (4,828 km) along the US East Coast and the coast of the Gulf of Mexico. Man-made canals connect rivers, **lagoons**, bays, and sounds.

The Intracoastal Waterway has two parts: the Atlantic Intracoastal Waterway and the Gulf Intracoastal Waterway. A connecting link was never completed.

FUN FACT: 19

THE INTRACOASTAL WATERWAY PLAYED AN IMPORTANT PART DURING WORLD WAR II.

Ships transporting goods along the Atlantic Coast were at risk during World War II (1939–1945). People feared enemy boats. The Intracoastal Waterway provided a safe inland route to eastern American ports.

25

SHIFTING WATERS

FUN FACT: 20

ENGINEERS CREATED A SEAWAY TO FIGHT EROSION ALONG AUSTRALIA'S EASTERN COAST.

In this photo, taken after the area was stabilized, or made stable, you can see where the Gold Coast Seaway meets the Pacific Ocean.

Boats sailing into eastern Australia once used the mouth of the Nerang River. However, winds changed its position and how deep it was. The moving waters even destroyed a town! Engineers made the entrance stable. Today, it's called the Gold Coast Seaway.

ENGINEERS PROBLEM-SOLVE!

PROBLEMS:

EROSION BY WIND

EROSION BY SAND

CHANGING SEAWAY ENTRANCE

SOLUTIONS:

BUILD BARRIERS CALLED BREAKWATERS (TO PROTECT AGAINST WAVES)

DREDGE DEEPER CHANNELS (TO STABILIZE THE SEAWAY ENTRANCE)

BUILD WITH **CONCRETE** AND ROCK (TO STABILIZE THE SEAWAY ENTRANCE)

Stabilizing the entrance to the Pacific Ocean was a necessary step before creating the Gold Coast Seaway, a waterway into eastern Australia's Gold Coast.

ENGINEERS CHANGE THE WORLD!

The waterways in this book have changed the world. They make it easier for us to travel and receive and send goods. The engineers who make waterways have a lot to consider, such as erosion, communities, and wildlife that live on the land and in the water.

From the ancient people who built the Grand Canal in China to those building and reshaping man-made waterways today, engineers shape our ways of life. Think about that next time you travel near a waterway—or on one!

In some places such as Bangkok, Thailand, visitors can see how canals continue to be a part of everyday life!

GLOSSARY

civilization: organized society with written records and laws

complex: having to do with something with many parts that work together

concrete: a hard, strong matter used for building and made by mixing cement, sand, and broken rocks with water

economy: the money made in an area and how it is made

engineering: using science and math to build better objects

erosion: the act of wearing away by water, wind, or ice

irrigation: the watering of a dry area by man-made means in order to grow plants

lagoon: an area of sea water separated from the ocean by rocks or sand

lock: a device for raising and lowering ships between stretches of water that are different levels

marathon: a race in which people run about 26 miles (42 km)

pyramid: a 3-D shape that has a rectangular base and triangular sides

technology: tools, machines, or ways to do things that use the latest discoveries to fix problems or meet needs

transportation: the act or process of moving people or things from one place to another

FOR MORE INFORMATION

BOOKS

Hinman, Bonnie. *Infrastructure of America's Inland Waterways*. Hallandale, FL: Mitchell Lane Publishers, 2018.

Pascal, Janet B. *What Is the Panama Canal?* New York, NY: Grosset & Dunlap, 2014.

Polinsky, Paige V. *Canals*. Minneapolis, MN: Abdo Publishing, 2018.

WEBSITES

Erie Canal

www.ducksters.com/history/us_1800s/erie_canal.php

Learn more about the Erie Canal and its amazing history.

How It Works

www.pancanal.com/eng/general/howitworks/index.html

Find out more about the famous Panama Canal.

INDEX